YOUR KNOWLEDGE HAS VALUE

- We will publish your bachelor's and master's thesis, essays and papers

- Your own eBook and book - sold worldwide in all relevant shops

- Earn money with each sale

Upload your text at www.GRIN.com and publish for free

Bibliographic information published by the German National Library:

The German National Library lists this publication in the National Bibliography; detailed bibliographic data are available on the Internet at http://dnb.dnb.de .

Imprint:

Copyright © 2009 GRIN Verlag, Open Publishing GmbH
Print and binding: Books on Demand GmbH, Norderstedt Germany
ISBN: 978-3-640-98270-7

This book at GRIN:

http://www.grin.com/en/e-book/176816/inverse-estimation-of-heat-flux-and-temperature-in-3d-finite-domain

Nauman Malik Muhamad

Inverse Estimation of Heat Flux and Temperature in 3D Finite Domain

GRIN Publishing

Inverse Estimation of Heat Flux and Temperature Distribution in 3D Finite Domain

Nauman Malik Muhammad

A thesis submitted in partial fulfillment of the requirement for the degree of Master of Science

2009. 06

This thesis has been examined and approved.

2009.06
....
Date

Department of Nuclear and Energy Engineering

GRADUATE SCHOOL

JEJU NATIONAL UNIVERSITY

Dedicated to

My parents, Saba

and the coming one

ACKNOWLEDGEMENTS

Start with the name of Almighty Allah, the most merciful, the most beneficent. First of all I would present my humble gratitude in front of Allah Who enabled me to accomplish the dignified cause of education and learning and I would pray to Him that He would make me able to utilize my knowledge and edification in the betterment of humanity and its development. It is a matter of great honor for me to extend my heartiest indebtedness to all those who have provided guidance to me and proffered a helping hand to me and have contributed directly or indirectly in this work. I feel extremely indebted to my advisor during my MS studies, Prof. Dr. Kim Sin for his continuous guidance and support throughout this period at Jeju National University. His thorough professionalism, thought-provoking lectures, seminars and discussions and a never-ending teaching appetite I will admire and cherish forever. His encouragement and guidance during my high time and support and kindness during the down times have all been a source of motivation towards learning more and more and giving out my best which at the least I have tried to during my MS studies.

I will always respect and honor Prof. Dr. Kim Kyung Youn for his helpful supervision and support during my studies and research as well. His all-round personality and insightful comments on my work has made me think and it has been a cause of constant motivation towards my research. I also wish to extend my deep-hearted thanks to Prof. Chang Bum Jin and other professors of my department especially, Prof. Won-Gi Chan, Prof. Lee Han Joo, and Prof. Lee Yoon Joon for their extremely useful and thoughtful lectures and seminars. I would always commend the devotion and hard work of Ms. Kim Mee Jan for her Korean language teachings. I would like to take this opportunity to thank Prof. El-Huan Kim and Prof. Kyung Hyun Choi for their kindness and encouragement.

There have been many friends over this period who have been constantly mentored me and have been a source of comfort and companionship and they never let me feel alone even in the times of worries and hardships. I would first like to thank Dr. Umer Zeeshan Ijaz (Cambridge University, UK) for whom I am greatly indebted for his endless help when he acted as a counselor as well as a friend at the same time. I am especially grateful to Shafqat-ur-Rehman (INRIA-France) and Anil Kumar Khambampati for all their help, support and friendship. The companionship I have enjoyed with Ahmar Rashid, Ahsan Rehman, Asif Ali Rehmani and Khalid Rehman and the sense of being living in a family is not very common these days and is an asset of my life and I hope this will continue in the future as well. The help and attention given to my wife during her illness by Mrs. Ahmar, Mrs. Ahsan, Mrs. Asif and Mrs. Khalid is unprecedented and we are greatly indebted towards the latter for her apt and self-motivated facilitation.

I would like to thank Lee Jeong-Seong, Ko Min-Seok, Lee Kyung Hyun, Adnan Ali, Salim Khan, Shafqat-ullah, Santosh Kumar, Srikant Saini, Guna, Anji Reddy, Iskandar Makhmudov, Muhammad Rakib, Abhijit Saha, Nadun, Mahanama and all JISO members for their friendship and support. I would especially thank Wang Rong Li for her assistance in the preparation of this thesis.

I am hugely indebted to the help extended to me by Mr. Subhan Gul, Mr. Abid Hussain, Mr. Asif Waseem, Mr. Kamran Chughtai, Mr. Sohail Sarwar and Khurram Kafeel at DRD-PAEC. I am especially gratified to Mr. Abid Hussain for his selfless and continuous support for me in Pakistan. He never let me feel that I am away from my homeland and always encouraged and guided me whenever I needed any advice. I would also express appreciation for the role of Pakistan Atomic Energy Commission, BK-21 and KOSEF for supporting during my MS studies.

It is an honor for me to pay reverence and salute to my parents who have taken and are still taking great care of me and have been a great source of encouragement for me. They have given me confidence, self-respect, patience and faith in the hard work. Finally, I would like to thank my wife Saba Nauman, for standing by me through all the thick and thin and stimulating the sense of confidence in me. She has always been a source of motivation, character and courage to me and I would always admire her patience, fortitude and determination.

Contents

List of Tables

요약

역전도 문제는 열 유속과 열을 물질로 전도하는 층의 온도를 측정하기 곤란한 많은 이론적이고 실용적인 응용분야에서 발생한다. 이에 따라 그러한 문제들을 분류하고 역으로 열 유속을 추정하기 위한 기법을 개발하는 것은 필수적인 과제가 되고 있다. Bayesian estimation technique 은 역 입력(inverse input)과 선택적으로 가중된 칼만 필터(adaptively weighted Kalman filter)의 열(bank)로 구성되는 state estimator 를 만들어 준다. Adaptive state estimator 는 이런 Bayesian estimation technique 과 semi-Markovian concept 을 조합 함으로서 얻을 수 있는 기법 중 하나이다. 본 연구는 정육면체의 접근 가능한 면(accessible faces)에서 온도가 측정되는 동안 한 쪽 끝에서 열이 주입되고 다른 모든 면들은 단열된 3 차원 열전도 문제를 다룬다. 이런 접근 가능한 면(accessible faces)에서 측정된 온도는 추정 알고리즘(estimation algorithm)에 입력되고 입력 열 유속(input heat flux)과 시스템의 각 점들에 대한 온도가 계산된다.

추정 알고리즘(estimation algorithm)의 robustness 와 실제 응용분야에서의 사용가능성을 검증하기 위해 사인파형의 열 유속(sinusoidal input flux)과 성형적으로 변하는 사각형 조합(combination of rectangular)등 다양한 입력 열 유속 시나리오를 시험한다. 또한 이러한 입력에서 estimator 의 성능 한계를 시험하고, 추정 알고리즘의 실제적인 응용가능성을 평가하기 위해 각 입력과 관련된 오차(error)도 비교하였다. 센서의 수와 위치의 중요성을 확인하기 위해 다른 센서의 수와 위치를 다르게 하여 시험한다. 많은 측정 센서를 설치하는 것은 비경제적이고 비효율적인 과정이므로 최적의 센서 수와 위치를 결정하는 것은 중요하다.

관용적인 오차 한계(tolerable error limitation)와 동시에 계산적으로 복잡하지 않은 선에서 최적의 격자크기(mesh size)를 얻기 위해 역 추정(inverse estimation)과정 전에 시스템의 지배방정식의 유한요소계산에 대한 종합적인 격자 민감도 분석(mesh sensitivity analysis)을 수행한다.

SUMMARY

Inverse heat conduction problems occur in many theoretical and practical applications where it is difficult or practically impossible to measure the heat flux generated and the temperature of the layer conducting the heat flux to the body. Thus it becomes imperative to devise some means to cater for such a problem and estimate the heat flux inversely. Adaptive state estimator is one such technique which works by incorporating the semi-Markovian concept into a Bayesian estimation technique thereby developing an inverse input and state estimator consisting of a bank of parallel adaptively weighted Kalman filters. The problem presented in this study deals with a three dimensional system of a cube with one end conducting heat flux and all the other sides are insulated while the temperatures are measured on the accessible faces of the cube. The measurements taken on these accessible faces are fed into the estimation algorithm and the input heat flux and the temperature distribution at each point in the system is calculated.

A variety of input heat flux scenarios have been examined to underwrite the robustness of the estimation algorithm and hence insure its usability in practical applications. These include sinusoidal input flux, a combination of rectangular, linearly changing and sinusoidal input flux and finally a step changing input flux. The estimator's performance limitations have been examined in these input set-ups and error associated with each set-up is compared to conclude the realistic application of the estimation algorithm in such scenarios. Different sensor arrangements, that is different sensor numbers and their locations are also examined to impress upon the importance of number of measurements and their location i.e. close or farther from the input area. Since practically it is both economically and physically tedious to install more number of measurement sensors, hence optimized number and location is very important to determine for making the study more application oriented.

Before the inverse estimation, a comprehensive mesh sensitivity analysis is given for the system's governing equation finite difference calculations to get an optimized mesh size for the forward and inverse analysis to be correct and within the tolerable error limits and computationally undemanding at the same time.

INTRODUCTION

1. Background – Inverse Problems

Inverse problems in engineering disciplines have always been an alluring area of interest for the researchers as it considerably simplifies the system identification and on the other hand is sufficiently accurate in estimating the parameters which are very difficult or impossible to measure due to system complexity, intricate geometry or many other technological or financial constraints and hence obtained data is the only source of obtaining the model parameters. A very simple representation of the inverse problem can be given as in Fig. 1 below.

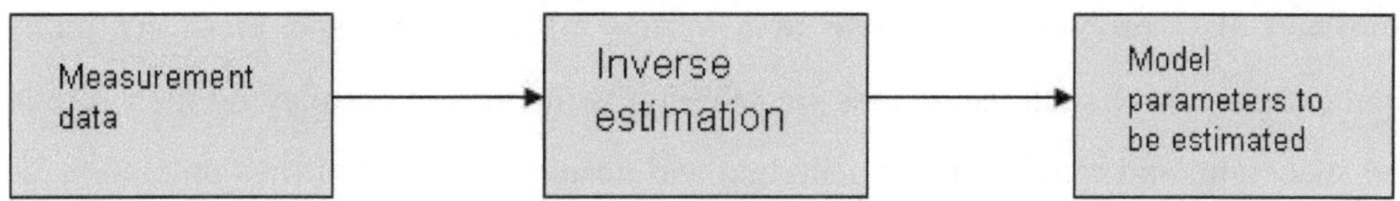

Fig. 1 A representation of the general inverse problem solution

The transformation from data to model parameters is a result of the interaction of a *physical system*, e.g., the Earth, the atmosphere, gravity etc. Inverse problems arise for example in heat conduction problems, geophysics, medical imaging (such as electrical impedance/resistance/capacitance tomography), remote sensing, ocean acoustic tomography, nondestructive testing, and astronomy.

Inverse problems are typically ill posed, as opposed to the well-posed problems more typical when modeling physical situations where the model parameters or material properties are known. Of the three conditions for a well-posed problem, i.e. existence, uniqueness and stability of the solution or solutions, the condition of stability is most often violated. In the sense of functional analysis, the inverse problem is represented by a mapping between metric spaces. While inverse problems are often formulated in infinite dimensional spaces, limitations to a finite number of measurements, and the practical consideration of recovering only a finite number of unknown parameters, may lead to the problems being recast in discrete form. In this case the inverse problem will typically be *ill-conditioned.* In these cases, regularization may be used to introduce mild assumptions on the solution and prevent overfitting.

3

Out of the various inverse problems involved in engineering disciplines, one of the most important is that of the inverse heat conduction problem, IHCP. It has been widely used in practical engineering problems involving the estimation of surface conditions or initial conditions as well as thermal properties of a body from known information like temperatures measured at the prescribed positions. It is sometimes necessary to calculate the transient surface heat flux and the surface temperature from a temperature measured at some location inside or outside the body. For example the case of a gun-barrel or nuclear reactor where heat flux is generated at a point which is virtually un-accessible and only the surface temperature can be measured. Keeping in view of the geometry of the body this IHCP problem can be solved using the inverse estimation techniques; however these problems are known to be severely ill-posed. Similarly IHC problems can arise in a variety of situations and there are many applications where such techniques are working wonders and are playing their role in the successful and reliable instrumentation and designing. Some of the common areas where IHC problems occur are as follows:

- Heat production and dissipation in micro-electronics where micro- devices produce heat and are covered with fins and other devices in contact.

- Heat production in nuclear reactors where the fuel is clothed in many layers of cladding and it is impossible to reach to the fuel surface.

- Designing of heat resistant materials and hence canopies of jets and rockets where it is practically impossible to install sensors on the inside and only outer surface temperatures can be measured.

- Determination of the heat transfer coefficient and outer surface conditions in the re-entry of a space vehicle.

- Designing of heat resistant gun-barrels.

- Designing of refractories and furnaces.

- Air-conditioning and refrigeration.

2.   Related Work

Although a lot of work has been carried by different researchers in the field of IHCP, this work is usually confined to 2D problems only and we do not see much of the work carried out for a 3D domain involving IHCP. Many techniques and inverse algorithms have been proposed, applied and inspected by various investigators in the field of inverse problems in engineering. There exist many methods to solve the IHCP and the majority of researchers use the approaches where the unknowns are determined to minimize the sum of squares of the differences between the measured and the computed temperatures at the selected spatial and/or temporal points. In general, the approaches adopt the iterative scheme and the regularizations are implemented to mitigate the ill-posedness of IHCP. The Kalman filter estimation method is successfully used in predicting multidimensional and nonlinear IHCP. Zheng and Murio [1] proposed a stable algorithm for 3D IHCP problem on a slab and then they [2] developed a numerical solution for the three-dimensional inverse heat conduction problem on a finite cube by applying a mollification procedure. A fully explicit space-marching finite-difference scheme was developed and numerical simulations were provided with excellent accuracy between estimated and exact solutions. Scarpa and Milano [3] estimated the time-dependent surface heat flux at one boundary of a one-dimensional system by using the Kalman smoothing technique, given the initial temperature distribution and the time-temperature history at an interior location. The numerical results show appreciable performance of the proposed technique, which provides a comprehensive way for using future temperature measurements. Kaipio and Somersalo [4] used a regularized version of Kalman filter and inverse boundary value problem was considered for a nonstationary object, i.e. those properties of the object that one seeks to estimate change as a function of time during the measurement sequence. In recent years, many applications appeared in which Kalman filter has been used in conjunction with recursive-least square algorithm (RLSA), for example the work of Tuan et al. [5–12], deals with one-dimensional and two-dimensional problems. Extending that work, recently, Jang et al. [13] has attempted to use a RLSA based on the Kalman filter to estimate the boundary heat flux varying impulsively with time by employing the finite-element scheme to discretize the problem in space, allowing multidimensional problems of various geometries to be treated.

Kalman [14] filter has been the governing filter of most of the techniques proposed in these studies, however for the higher dimension problems, the straight forward implementation of the Kalman filter becomes difficult as the size of covariance equation increases. Therefore, one of the most important prerequisites for the successful implementation of a Kalman filter for the purpose of real-time estimation is the development of a reliable low dimensional model, hence, dimension reduction techniques like Karhunen-Love Galerkin procedure is used with Kalman filter by Park and Jung [15] for solving multidimensional heat conduction problems. For the case of nonlinear IHCP, extended version of discrete Kalman filter have been used by Daouas and Radhouani [16, 17] to nonlinear IHCP to estimate surface heat flux density. Huang and Tsai [18] solved the IHCP using the conjugate gradient method. Boundary Element Method (BEM)-based inverse algorithm utilizing the iterative regularization method was successfully used to solve the Inverse Heat Conduction Problem (IHCP) for estimating the unknown transient boundary heat flux in a 2D domain with different arbitrary geometries.

An important parameter in solving the inverse problems is that of stability because of the prohibitive ill-posedness of the problem. As the IHCP finds wide applications in many thermal-related industries, it is of great practical importance to study the various effects on the stability of the inverse solutions. Surprisingly, despite so many existing inverse techniques, a systematic study of the stability of the inverse solutions has not been pursued by many researchers. Most of the techniques do not give a quantitative method for determining the computed errors due to noise in temperature measurements. The reason for a lack of studies on the stability of the solution of the inverse problem is simple. The IHCP is already very difficult and its solution instability analysis is even more. One such study has been carried out by Ling and Atluri [19]. Two matrix algebraic tools are provided for studying the solution-stabilities of inverse heat conduction problems. The propagations of the computed temperature errors, as caused by noise in temperature measurements are given and the spectral norm analysis reflects the effect of the computational time steps, the sensor locations and the number of future temperatures on the computed error levels. Hsu [20] presented an account of the inverse estimation of the boundary conditions in a 3D inverse hyperbolic heat conduction problem. Finite-difference methods are employed to discretize the problem domain, and then a linear inverse model is constructed to identify the unknown boundary condition.

It is noteworthy that Tuan et al. [5] developed the RLSA based on the Kalman filter for two-dimensional IHCP to estimate the boundary heat flux varying impulsively with time. Their approach gives good estimates for estimating unknown heat sources or heat flux inputs on the boundaries. They have proposed improvement in Kalman filter with RLSA approach by having RLSA weighted by forgetting factor to robustly extract the unknowns. The maximum likelihood type estimator combined with Huber psi-function is used to construct the weighting forgetting factor. In the context of forgetting factor, Wang et al. [21] proposed extended Kalman filter with RLSA weighted by forgetting factor to estimate nonlinear heat conduction problems.

3.   Kalman Filtering and Present Problem

Right from its inception in 1959 by R. E. Kalman [14], there have been many modifications, additions and developments proposed in the Kalman filter and it has been proved as a building block for many revolutionary estimation techniques. The modifications to Kalman filter are in hundreds of thousands and the break-through which it has provided in the field of numerical analysis is priceless and far-reaching as well. The main contribution of Kalman filter is in the field of inverse problems and state estimation in particular. Thereby deploying the Kalman filter, Moose et al. [22] proposed an Adaptive State Estimator (ASE) for passive underwater tracking of maneuvering targets. The state estimator is designed specifically for a system containing independent unknown or randomly switching input and measurement biases. In modeling the stochastic system, it is assumed that the bias sequence dynamics for both input and measurement can be modeled by a semi-Markov process. By incorporating the semi-Markovian concept into a Bayesian estimation technique, an estimator consisting of a bank of parallel adaptively weighted Kalman filters was developed. Despite the large and randomly varying biases, the proposed estimator provides a reasonable estimate of the system states. The Bayesian computational technique has many advantages as it is able to quantify system uncertainty and random data error, to derive a probabilistic description of the inverse solution, to provide extensive spatial/temporal regularization to the ill-posedness of the inverse problem, and to allow adaptive sequential estimation. Wang and Zabaras in [23–26] developed a computational framework that integrates computational mathematics,

Bayesian statistics, statistical computation, and reduced-order modeling to address data-driven inverse heat and mass transfer problems.

In the context of above mentioned filtering and estimation technique the present work deals with the input heat flux and the temperature distribution in a 3D heat conduction domain using the ASE based on Kalman filter. It is worth mentioning here that Kim et al. [27] deployed ASE to one-dimensional IHCP for estimation of input heat flux and then carrying their work forward and considering measurement bias into account, Ijaz et al. [28] have focused their research on a typical 2D heat conduction problem. Their study shows that ASE consisting of Kalman filters connected in parallel gives good performance in the presence of measurement bias also. Each filter has its operating bound limiting the range of the unknown input heat flux and measurement bias.

Traditionally, inverse problems are divided into two sequential stages: analysis and optimization. In the analysis stage, the values of unknown states are initially assumed, and then a numerical method (e.g. the finite-difference method or the finite-element method) is used to obtain the exact solution. In the optimization stage, the measurements data are compared with the results predicted in the previous stage, and are then compounded to form a non-linear problem. An optimization and filtering process is then employed to derive the optimal estimated solution. We have used the finite difference method for the discretization of the domain and a 4th-order Runge-Kutta method is deployed to get the numerical solution of the problem. ASE is deployed then for the estimation of input flux used in the previous stage to get the numerical solution and then state estimation i.e. temperature distribution estimation is obtained finally. Different types of boundary condition are used for the verification of results and different number of sensors and sensor arrangements are also analyzed to get the knowledge of estimator's dependence on measurement numbers and measurement locations. A comprehensive mesh sensitivity analysis is provided to cater for the inaccuracy and error involved in the estimation.

II.   Forward Problem

1.   Problem Description and Finite Difference Solution

Let us consider a three-dimensional cube, initially at temperature $\overline{T}(\overline{x},\overline{y},\overline{z},0) = 0$. For times $\overline{t} > 0$, all the faces of the cube are kept insulated except the face $(\overline{x},\overline{y},\overline{z}) = (\overline{L}, 0 \leq \overline{y} \leq \overline{W}, 0 \leq \overline{z} \leq \overline{H})$ which is conducting a heat flux $\overline{q}(\overline{t})$ to the cube. Fig. 1 illustrates the heat conduction problem considered along with boundary and initial conditions. The governing equation of this problem in dimensional form is given as

$$\frac{\partial \overline{T}(\overline{x},\overline{y},\overline{z},\overline{t})}{\partial \overline{t}} = \alpha \left(\frac{\partial^2 \overline{T}(\overline{x},\overline{y},\overline{z},\overline{t})}{\partial \overline{x}^2} + \frac{\partial^2 \overline{T}(\overline{x},\overline{y},\overline{z},\overline{t})}{\partial \overline{y}^2} + \frac{\partial^2 \overline{T}(\overline{x},\overline{y},\overline{z},\overline{t})}{\partial \overline{z}^2} \right), \tag{1}$$

$$0 \leq \overline{x} \leq \overline{L}, 0 \leq \overline{y} \leq \overline{W}, 0 \leq \overline{z} \leq \overline{H}, \overline{t} > 0$$

where $\overline{x},\overline{y},\overline{z}$ and $\overline{t}$ are space and time coordinates respectively, α is the thermal diffusivity and $\overline{L}, \overline{W}$ and $\overline{H}$ are the length, width and height of the cube respectively.

If some non-dimensional parameters are defined as

$$x = \frac{\overline{x}}{\overline{L}}, \quad y = \frac{\overline{y}}{\overline{L}}, \quad z = \frac{\overline{z}}{\overline{L}}, \quad t = \frac{\alpha \overline{t}}{\overline{L}^2}, \quad T = \frac{\overline{T} - \overline{T}_{in}}{\overline{T}_o - \overline{T}_{in}}$$

and substituted in Eq. (1), then the dimensionless form of Eq. (1) is given as:

$$\frac{\partial T(x,y,z,t)}{\partial t} = \frac{\partial^2 T(x,y,z,t)}{\partial x^2} + \frac{\partial^2 T(x,y,z,t)}{\partial y^2} + \frac{\partial^2 T(x,y,z,t)}{\partial z^2}, \tag{2}$$

$$0 \leq x \leq 1, 0 \leq y \leq 1, 0 \leq z \leq 1, t > 0$$

with the following boundary and initial condition

$$\frac{\partial T(0,y,z,t)}{\partial x} = 0, \ 0 \leq y \leq 1, 0 \leq z \leq 1, t > 0 \tag{3}$$

$$\frac{\partial T(1,y,z,t)}{\partial x} = q(t), \ 0 \leq y \leq 1, 0 \leq z \leq 1, t > 0 \tag{4}$$

$$\frac{\partial T(x,0,z,t)}{\partial y} = 0, \ 0 \leq x \leq 1, 0 \leq z \leq 1, t > 0 \tag{5}$$

$$\frac{\partial C(x,1,z,t)}{\partial y} = 0, \ 0 \leq x \leq 1, 0 \leq z \leq 1, t > 0 \tag{6}$$

$$\frac{\partial T(x,y,0,t)}{\partial z} = 0, \ 0 \le x \le 1, 0 \le y \le 1, t > 0 \tag{7}$$

$$\frac{\partial T(x,y,1,t)}{\partial z} = 0, \ 0 \le x \le 1, 0 \le y \le 1, t > 0 \tag{8}$$

$$T(x,y,z,0) = 0, \ \ 0 \le x \le 1, 0 \le y \le 1, 0 \le z \le 1 \tag{9}$$

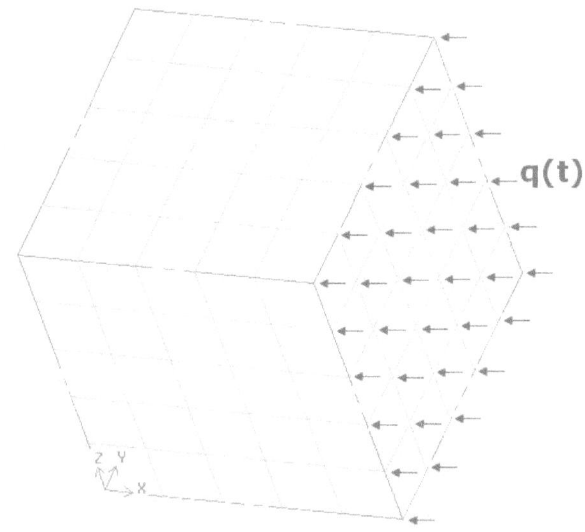

Fig. 2 Meshed representative diagram of the 3D cube

In the above equations, T is the dimensionless temperature and x, y, z and t are the dimensionless space and time coordinates, respectively while q is the input heat flux.

Now applying the Central Finite Difference method on Eq. (2) implies that

$$\dot{T}_{i,j,k} = [\frac{1}{\Delta x^2}]T_{i+1,j,k} + [\frac{1}{\Delta y^2}]T_{i,j+1,k} + [\frac{1}{\Delta z^2}]T_{i,j,k+1} - 2[\frac{1}{\Delta x^2} + \frac{1}{\Delta y^2} + \frac{1}{\Delta z^2}]T_{i,j,k} +$$
$$[\frac{1}{\Delta x^2}]T_{i-1,j,k} + [\frac{1}{\Delta y^2}]T_{i,j-1,k} + [\frac{1}{\Delta z^2}]T_{i,j,k-1} \tag{10}$$

for $i = 1,2,3,\ldots\ldots,M$, $j = 1,2,3,\ldots\ldots\ldots,N,$ and $k = 1,2,3,\ldots\ldots\ldots,O$, and $M,N,O > 1$, where M, N and O are the total number of spatial nodes for x, y, z directions, respectively, and $\Delta x = 1/(M-1)$, $\Delta y = 1/(N-1)$ and $\Delta z = 1/(O-1)$ are the space intervals. Application of the Central Finite Difference method on the boundary conditions, Eqs. (3) through (8) gives

$$T_{0,j,k} = T_{2,j,k}, \tag{11}$$

$$T_{M+1,j,k} = T_{M-1,j,k} + 2\Delta x q(t), \tag{12}$$

$$T_{i,0,k} = T_{i,2,k}, \tag{13}$$

$$T_{i,N+1,k} = T_{i,N-1,k}, \tag{14}$$

$$T_{i,j,0} = T_{i,j,2}, \tag{15}$$

and

$$T_{i,j,O+1} = T_{i,j,O-1}. \tag{16}$$

From Eq. (10)

$$a = \frac{1}{\Delta x^2}, \quad b = \frac{1}{\Delta y^2}, \quad c = \frac{1}{\Delta z^2} \quad \text{and} \quad d = -2\left[\frac{1}{\Delta x^2} + \frac{1}{\Delta y^2} + \frac{1}{\Delta z^2}\right].$$

Therefore Eq. (10) can be re-written as

$$\dot{T} = aT_{i+1,j,k} + bT_{i,j+1,k} + cT_{i,j,k+1} + dT_{i,j,k} + aT_{i-1,j,k} + bT_{i,j-1,k} + cT_{i,j,k-1} \tag{17}$$

From Eqs. (12) through (17) and associated with fictitious process noise inputs [29], the continuous-time state equation can be written as

$$\dot{T}(t) = \Psi T(t) + \Omega[q(t) + w(t)] \tag{18}$$

here

$$T(t) = \begin{bmatrix} T_{1,1,1} & T_{1,1,2} & \cdots & T_{1,1,O} & T_{2,1,1} & T_{2,1,2} & \cdots & T_{2,1,O} & \cdots & T_{M,1,O} & T_{1,2,1} & T_{1,2,2} & \cdots & T_{M,2,O} & \cdots & T_{M,N,O} \end{bmatrix}^T \tag{19}$$

and all $T_{i,j,k}$ are continuous w.r.t. time t.

The coefficient matrix $\Psi \in \mathbb{R}^{(MNO)\times(MNO)}$ is given by

$$\Psi = \begin{bmatrix} \Psi_1 & 2\Upsilon & \varnothing & \varnothing & \varnothing & & \cdots & & \varnothing \\ \Upsilon & \Psi_1 & \Upsilon & \varnothing & \varnothing & & \cdots & & \varnothing \\ \varnothing & \Upsilon & \Psi_1 & \Upsilon & \varnothing & & \cdots & & \varnothing \\ \vdots & & \ddots & \ddots & \ddots & \ddots & \ddots & & \vdots \\ \varnothing & & \cdots & & \varnothing & \Upsilon & \Psi_1 & \Upsilon & \varnothing \\ \varnothing & & \cdots & & \varnothing & \varnothing & \Upsilon & \Psi_1 & \Upsilon \\ \varnothing & & \cdots & & \varnothing & \varnothing & \varnothing & 2\Upsilon & \Psi_1 \end{bmatrix}$$

where

$$\Psi_1 = \begin{bmatrix} A & C & \Theta & \Theta & \Theta & & \cdots & & \Theta \\ B & A & B & \Theta & \Theta & & \cdots & & \Theta \\ \Theta & B & A & B & \Theta & & \cdots & & \Theta \\ \vdots & & \ddots & \ddots & \ddots & \ddots & \ddots & & \vdots \\ \Theta & & \cdots & & \Theta & B & A & B & \Theta \\ \Theta & & \cdots & & \Theta & \Theta & B & A & B \\ \Theta & & \cdots & & \Theta & \Theta & \Theta & C & A \end{bmatrix} \quad \text{and}$$

$$\Upsilon = \begin{bmatrix} D & \Theta & \Theta & \Theta & \Theta & & \cdots & & \Theta \\ \Theta & D & \Theta & \Theta & \Theta & & \cdots & & \Theta \\ \Theta & \Theta & D & \Theta & \Theta & & \cdots & & \Theta \\ \vdots & & \ddots & \ddots & \ddots & \ddots & \ddots & & \vdots \\ \Theta & & \cdots & & \Theta & \Theta & D & \Theta & \Theta \\ \Theta & & \cdots & & \Theta & \Theta & \Theta & D & \Theta \\ \Theta & & \cdots & & \Theta & \Theta & \Theta & \Theta & D \end{bmatrix}$$

In which

$$A = \begin{bmatrix} d & 2c & 0 & 0 & 0 & \cdots & 0 \\ c & d & c & 0 & 0 & \cdots & 0 \\ 0 & c & d & c & 0 & \cdots & 0 \\ \vdots & \ddots & \ddots & \ddots & \ddots & \ddots & \vdots \\ 0 & \cdots & 0 & c & d & c & 0 \\ 0 & \cdots & 0 & 0 & c & d & c \\ 0 & \cdots & 0 & 0 & 0 & 2c & d \end{bmatrix} \qquad B = \begin{bmatrix} a & 0 & 0 & 0 & 0 & \cdots & 0 \\ 0 & a & 0 & 0 & 0 & \cdots & 0 \\ 0 & 0 & a & 0 & 0 & \cdots & 0 \\ \vdots & \ddots & \ddots & \ddots & \ddots & \ddots & \vdots \\ 0 & \cdots & 0 & 0 & a & 0 & 0 \\ 0 & \cdots & 0 & 0 & 0 & a & 0 \\ 0 & \cdots & 0 & 0 & 0 & 0 & a \end{bmatrix}$$

$$C = 2B$$

$$D = \begin{bmatrix} b & 0 & 0 & 0 & 0 & \cdots & 0 \\ 0 & b & 0 & 0 & 0 & \cdots & 0 \\ 0 & 0 & b & 0 & 0 & \cdots & 0 \\ \vdots & \ddots & \ddots & \ddots & \ddots & \ddots & \vdots \\ 0 & \cdots & 0 & 0 & b & 0 & 0 \\ 0 & \cdots & 0 & 0 & 0 & b & 0 \\ 0 & \cdots & 0 & 0 & 0 & 0 & b \end{bmatrix}$$

The matrices $\Psi_1, \varnothing$ and $\Upsilon \in \mathbb{R}^{M \times O}$ while the sub-matrices A, B, C, D and Θ are $\mathbb{R}^{O \times O}$, where Θ is a null sub-matrix. The input matrix $\Omega \in \mathbb{R}^{(MNO) \times 1}$ is given by

$$\Omega = 2\Delta x \begin{bmatrix} \omega_{1,1,1} & \omega_{1,1,2} & \cdots & \omega_{1,1,O} & \omega_{2,1,1} & \omega_{2,1,2} & \cdots & \omega_{2,1,O} & \cdots & \omega_{M,1,O} & \omega_{1,2,1} & \omega_{1,2,2} & \cdots & \omega_{M,2,O} & \cdots & \omega_{M,N,O} \end{bmatrix}^T \tag{20}$$

and all $\omega_{i,j,k}$ are continuous w.r.t. time. The elements of Ω will be all zero except at the nodes where heat is being applied i.e. all $\omega_{M,j,k}$ will be 1 and all the other elements will be zero.

Eq. (18) is the reference equation for inverse estimation results. It should be solved numerically for the acquisition of true results with whom the inverse solution should be compared. Keeping in view the complexity of the problem involved we have chosen to solve the equation with 4th-order Runge-Kutta method. The general algorithm of Runge-Kutta method used [30] is given as follows:

$$T_{\tau+1} = T_\tau + \Delta t \left(\frac{1}{6} f(T_\tau, t_\tau) + \frac{1}{3} f(T^*_{\tau+\frac{1}{2}}, t_{\tau+\frac{1}{2}}) + \frac{1}{3} f(T^{**}_{\tau+\frac{1}{2}}, t_{\tau+\frac{1}{2}}) + \frac{1}{6} f(T^*_{\tau+1}, t_{\tau+1}) \right) \tag{21}$$

where

$$T^*_{\tau+\frac{1}{2}} = T_\tau + \frac{\Delta t}{2} f(T_\tau, t_\tau) \tag{22}$$

$$T^{**}_{\tau+\frac{1}{2}} = T_\tau + \frac{\Delta t}{2} f(T^*_{\tau+\frac{1}{2}}, t_{\tau+\frac{1}{2}}) \tag{23}$$

$$T^*_{\tau+1} = T_\tau + \Delta t \cdot f(T^{**}_{\tau+\frac{1}{2}}, t_{\tau+\frac{1}{2}}) \tag{24}$$

2.   Sensitivity Analysis

A very important aspect of numerical solutions is that of mesh sensitivity analysis both with respect to temporal and spatial meshes. The importance of choice of a reasonable time step and element numbers become all the more important when one needs to optimize the computational burden at one side and the accuracy requirements are stringent on the other side. Therefore a detailed mesh sensitivity analysis is the requirement before going further to the inverse estimation.

A constant heat flux value of 5 units is used on the right face of the cube in Fig. 2 for carrying out sensitivity analysis. Fig. 3 represents the temperature distribution at three points in the domain i.e. $(x, y, z) = (0,0,0)$, $(0.5, 0.5, 0.5)$ and $(1.0, 1.0, 1.0)$ with a time step size of 0.01 and a spatial mesh of 11 nodes in x, y and z axis respectively. It means that there are $11 \times 11 \times 11 = 1331$ mesh nodes in total in the cube. It is clear from the figure that after approximately 0.20 seconds, the trend becomes linear everywhere in the domain. Therefore it is of interest to analyze the non-linear temperature distribution portion of time. Fig. 4 represents the temperature distribution at $(x, y, z) = (1, 1, 1)$ for 0.20 seconds with $11 \times 11 \times 11$ mesh size. Fig. 5 represents the temperature distribution at the same points with different mesh sizes.

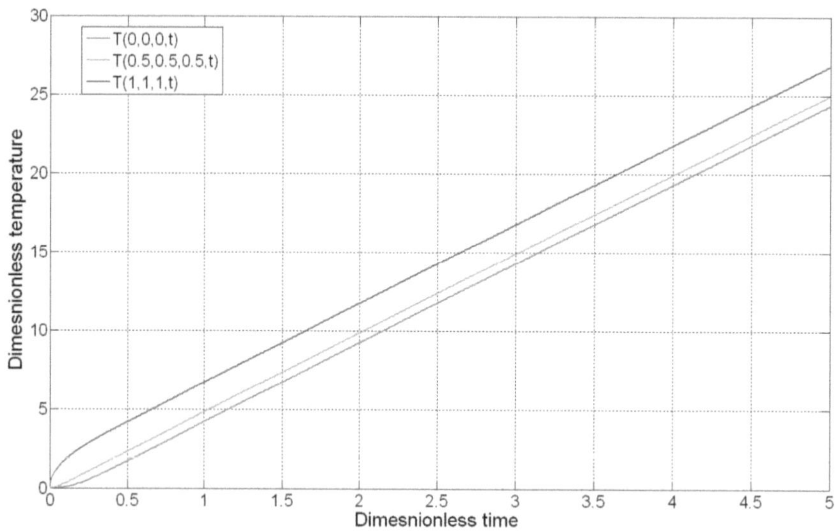

Fig. 3 Temperature distribution representation with $11 \times 11 \times 11$ at three pints in the cube

The objective of this analysis is the identification of a minimum mesh size which estimates this portion with reasonable accuracy. That mesh size will define the threshold that must at least be followed for acceptable numerical results. However we must have an authentic means with which we can compare the different mesh results and verify them. The best tool that we could have is the existence of analytical solution. Fortunately the problem we are dealing with can be considered as a 1D problem if we assume the material as homogenous and isotropic. In that case since all the faces other than the one conducting heat flux are insulated, therefore we can safely approximate the problem as 1D heat conduction problem. An analytical solution for such a problem is available [31] and is given as:

$$T(x,t) = \frac{qt}{\rho Cl} + \frac{ql}{K}\left\{ \frac{3x^2 - l^2}{6l^2} - \frac{2}{\pi^2}\sum_{n=1}^{\infty}\frac{(-1)^n}{n^2}e^{-Kn^2\pi^2 t/l^2}\cos\frac{n\pi x}{l} \right\} \tag{25}$$

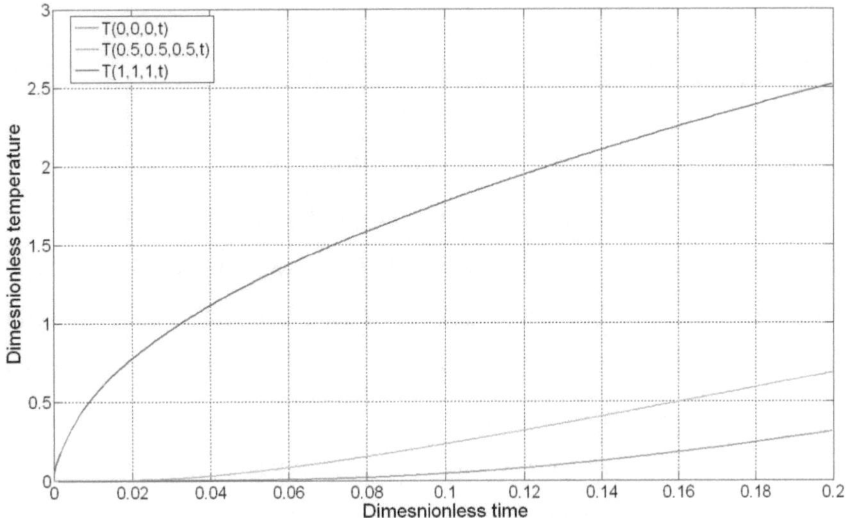

Fig. 4 Temperature distribution representation for non-linear portion of time at

$$(x, y, z) = (1, 1, 1)$$

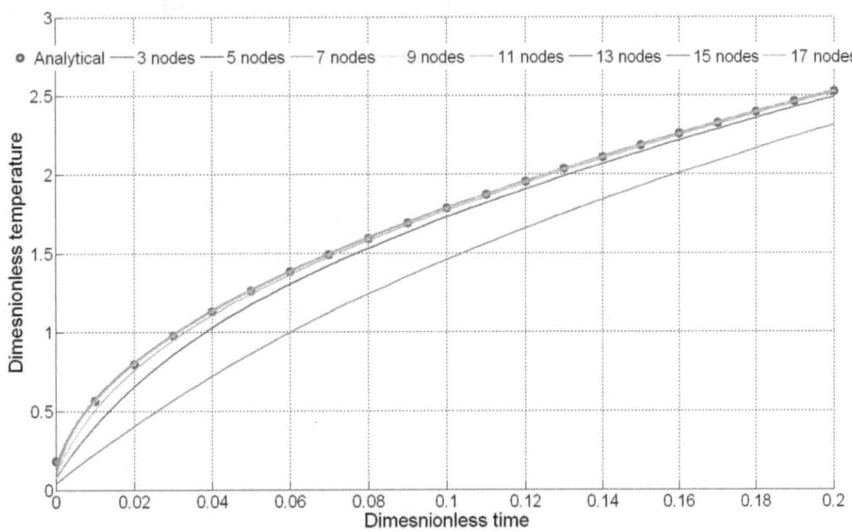

Fig. 5 Comparison of different mesh size results with analytical solution at $(x, y, z) = (1, 1, 1)$

where ρ is the density, C is the heat capacity, K the thermal conductivity and l the total length. To make the parameters consistent with our dimensionless model, we will take all of these parameters as unity and the solution will work for our problem.

Table 1 summarizes the relative error associated with each mesh size when compared to the above given analytical solution, where relative error is given as

$$R_E = \sqrt{\frac{\sum (T_{analytical} - T_{runge-kutta})^2}{\sum (T_{analytical})^2}} \tag{26}$$

As is evident from the table, when only 0.1 units time is analyzed, a coarse mesh gives a large error which goes on decreasing as the mesh size is reduced to finer values. However, as the mesh is made finer, the total computational time increases manifolds and it becomes practically useless to use a very fine mesh. An optimization compromise becomes therefore necessary and we have to choose a mesh size which gives sufficiently accurate results and is also computationally less tedious. Another aspect of the mesh-error analysis is that as the total analysis time is increased, even a coarse mesh gives appreciable results. The reason is that once the temperature distribution becomes linear w.r.t. time, the numerical solution starts working better than it does for a non-linear temperature distribution region. Hence the error is less when we increase the total analysis time. A reasonable mesh size can be one which gives less than 0.5% error when compared to the analytical solution. Hence making the same as our criteria $9 \times 9 \times 9$ elements is chosen as the mesh structure same will be used for the inverse estimation calculations. Although $7 \times 7 \times 7$ mesh will also give a reasonable result as the relative error involved with this mesh size is also less than 0.5%, but to be on the safe side and for the insurance of fairly accurate results, we will opt for a mesh finer than this one, that is the mesh with $9 \times 9 \times 9 = 729$ mesh nodes. By taking a finer mesh size will be asking for more computational complexity, but of course the results will be close to accurate and less prone to failure and slip-up which must be the priority.

Table 1 Percent relative error for different mesh sizes and different simulation times

No. of nodes	$3\times3\times3$	$5\times5\times5$	$7\times7\times7$	$9\times9\times9$	$11\times11\times11$	$13\times13\times13$	$15\times15\times15$	$17\times17\times17$
$t_f = 0.1$	9.17	0.93	0.17	0.043	0.013	0.0065	0.0054	0.0073
$t_f = 0.2$	3.46	0.27	0.048	0.011	0.0028	0.0016	0.0015	0.0022
$t_f = 1.0$	0.18	0.011	0.0018	0.0004	0.00014	0.00012	0.00015	0.00008
$t_f = 2.0$	0.041	0.0023	0.00033	0.000088	0.00006	0.00006	0.000087	0.00003
$t_f = 5.0$	0.0052	0.00022	0.000035	0.000035	0.00004	0.00053	0.000059	0.00002

III. INVERSE ESTIMATION

1. Kalman Filtering

In 1960, R.E. Kalman published his famous paper describing a recursive solution to the discrete data linear filtering problem [14]. Since that time, due in large part to advances in digital computing the Kalman filter has been the subject of extensive research and application, particularly in the area of autonomous or assisted navigation and inverse problems. The Kalman filter addresses the general problem of trying to estimate the state $x \in \mathbb{R}^n$ of a discrete-time controlled process that is governed by the linear stochastic difference equation,

$$x_\tau = Ax_{\tau-1} + Bu_{\tau-1} + w_{\tau-1} \tag{27}$$

with a measurement $z \in \mathbb{R}^m$ that is

$$z_\tau = Hx_\tau + v_\tau \tag{28}$$

where A and B are state transition and input matrix respectively, u the input bias, z the measurement sequence and H is the measurement matrix and τ is the discretized time index. The random variables w and v represent the process and measurement noise respectively.

The Kalman filter estimates a process by using a form of feedback control: the filter estimates the process state at some time and then obtains feedback in the form of (noisy) measurements. As such, the equations for the Kalman filter fall into two groups: *time update* equations and *measurement update* equations. The time update equations are responsible for projecting forward (in time) the current state and error covariance estimates to obtain the *a priori* estimates for the next time step. The measurement update equations are responsible for the feedback—i.e. for incorporating a new measurement into the *a priori* estimate to obtain an improved *a posteriori* estimate.

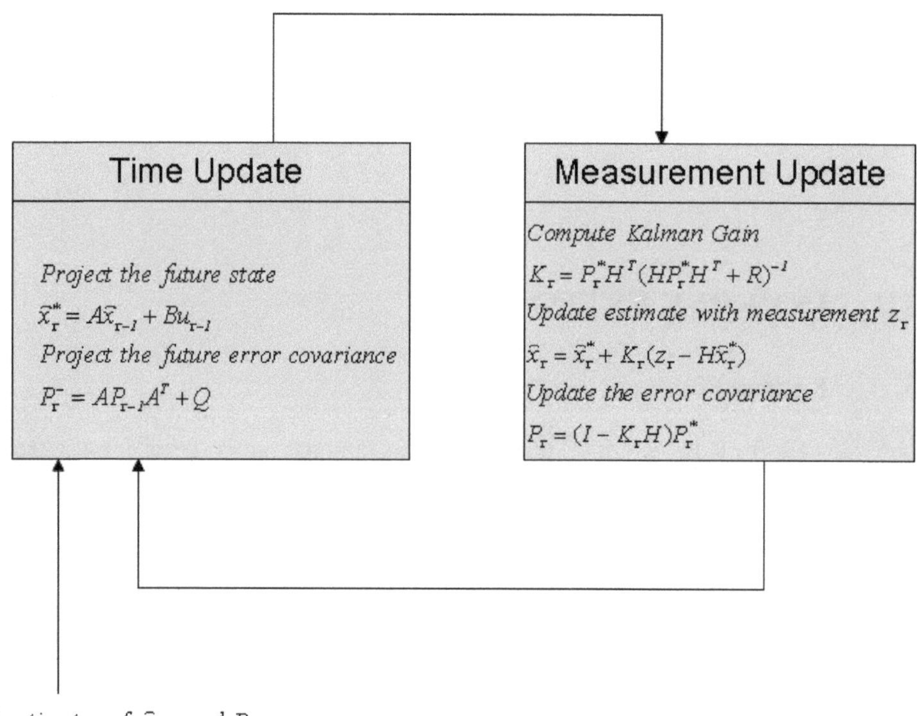

Fig. 6 The discrete Kalman filter cycle. The *time update* projects the current state estimate ahead in time. The *measurement update* adjusts the projected estimate by an actual measurement at that time

The time update equations can also be thought of as *predictor* equations, while the measurement update equations can be thought of as *corrector* equations. Indeed the final estimation algorithm resembles that of a *predictor-corrector* algorithm for solving numerical problems as shown in Fig. 6.

The first task during the measurement update is to compute the Kalman gain K. The next step is to actually measure the process to obtain the measurement z_k using the measurement matrix H, and then to generate an a posteriori state estimate by incorporating the measurement. The final step is to obtain an *a posteriori* error covariance estimate P_k. Q and R being coefficient of process noise and coefficient of measurement noise respectively. After each time and measurement update pair, the process is repeated with the previous *a posteriori* estimates used to project or predict the new *a priori* estimates. This recursive nature is one of the very appealing features of the Kalman filter as it makes practical implementations much more feasible than many other algorithms.

2.   Adaptive State Estimation - ASE

The Adaptive State Estimator, ASE, deals with the Bayesian state estimation technique which has been set up for the inverse problem to estimate the unknown input flux with white Gaussian noise associated in the state and measurement equations using a chain of parallely weighted Kalman filters. The input flux and bias terms are modeled as a semi-Markov process. The system analysis of semi-Markov processes is covered in detail by Howard [32]. A semi-Markov process, briefly, is the process that changes states according to a discrete time Markov Chain with finite number of states, and the time the system spends in each state is a random variable.

The state equation, Eq. (18), discretized over time intervals of length Δt is given by

$$X_{(\tau+1)} = \Phi X_{(\tau)} + \Gamma[q_{(\tau)} + w_{(\tau)}] \tag{29}$$

where

$$X_{(\tau)} = \begin{bmatrix} T_{1,1,1} & T_{1,1,2} & \cdots & T_{1,1,O} & T_{2,1,1} & T_{2,1,2} & \cdots & T_{2,1,O} & \cdots & T_{M,1,O} & T_{1,2,1} & T_{1,2,2} & \cdots & T_{M,2,O} & \cdots & T_{M,N,O} \end{bmatrix}^T \tag{30}$$

and all the $T_{i,j,k}$ are discretized w.r.t. τ.

$$\Phi = e^{\Psi \Delta t}, \tag{31}$$

$$\Gamma = \int_{\tau\Delta t}^{(\tau+1)\Delta t} \exp\{\Psi[(\tau+1)\Delta t - \theta]\}\Omega d\theta \tag{32}$$

Here $X_{(\tau)}$ represents the state vector, Φ is the state transition matrix, Γ the input matrix, $q_{(\tau)}$ the sequence of deterministic input, and $w_{(\tau)}$ the process noise vector, assumed to be zero mean and white Gaussian with variance $E\{w_{(\tau)}w^T{}_{(j)}\} = Q\delta_{\tau j}$ where Q is the coefficient of process noise and $\delta_{\tau j}$ is the Kronecker delta. The measurement equation in matrix form can be described as

$$Z_{(\tau+1)} = HX_{(\tau+1)} + v_{(\tau+1)} \tag{33}$$

where $Z_{(\tau)}$ is the observation vector and $v_{(\tau)}$ represents the measurement noise vector assumed to be zero mean and white noise. The variance of $v_{(\tau)}$ is given by $E\{v(\tau)v^T(j)\} = R\delta_{\tau j}$. Here, $R = \sigma^2 I$ where σ and I represent the standard deviation of measurement noise and identity matrix, respectively. $H \in \mathbb{R}^{1 \times (MNO)}$ is the measurement matrix given as

$$H = \begin{bmatrix} 0 & \cdots & 0 & 1 & 0 & \cdots & 0 & 1 & 0 & \cdots & 0 \end{bmatrix} \tag{34}$$

and the value of H is 1 at the measurement node. The state and measurement equations are given as Eqs. (29) and (33) simultaneously. An independent semi-Markov process governs the input $q_{(\tau)}$ and it can independently take on any of the possible discrete vectors $[q^{(1)} \, q^{(2)} \cdots q^{(n)}]$ for a random duration of time before a transition to a new bias occurs. Here n represents the possible discrete values of input flux. The range of vector $q^{(\alpha)}$, where $\alpha = 1,2,3,\cdots,n$, is modeled such that it spans the entire possible ranges of $q_{(\tau)}$. Since the adaptive estimator was developed by assuming that q can take on n possible discrete values, it is recognized that this could lead to excessive computation if n is large. Thus by assuming that q is uniformly distributed between adjacent vectors $q^{(\alpha)}$ and $q^{(\alpha+1)}$ [23] and incorporating this additional uncertainty as Q_b into the estimator gain, it becomes possible for the estimator to converge to an unbiased estimate of $x_{\tau+1}$. The optimal estimate of the state vector can be derived from the conditional mean by applying Bayesian conditional probability theory. The complete derivation of the adaptive state estimation and the background concept is given in [23] and the required state estimator equations are given below with the block diagram presented in Fig. 7.

$$X_{(\tau+1)} = \sum_{\alpha=1}^{n} \widehat{X}_{(\tau+1)}^{(\alpha)} W_{(\tau+1)}^{(\alpha)} \tag{35}$$

$$q_{(\tau)} = \sum_{\alpha=1}^{n} q^{(\alpha)} W_{(\tau+1)}^{(\alpha)} \tag{36}$$

where

$$\widehat{X}_{(\tau+1)}^{(\alpha)} = \Phi \widehat{X}_{(\tau)}^{(\alpha)} + \Gamma q^{(\alpha)} + K_{(\tau+1)}[Z_{(\tau+1)} - H\Phi \widehat{X}_{(\tau)}^{(\alpha)} - H\Gamma q^{(\alpha)}] \tag{37}$$

In the above equation

$$K_{(\tau+1)} = M_{(\tau+1)} H^{T} [HM_{(\tau+1)} H^{T} + R]^{-1} \tag{38}$$

$$M_{(\tau+1)} = \Phi P_{(\tau)} \Phi^{T} + \Gamma Q_{b} \Gamma^{T} + \Gamma Q \Gamma^{T} \tag{39}$$

$$P_{(\tau+1)} = [I - K_{(\tau+1)} H] M_{(\tau+1)} \tag{40}$$

The weight matrix W is defined as

$$W_{(\tau+1)}^{(\alpha)} = \beta W_{(\tau)}^{(\alpha)} + (1-\beta)\breve{W}_{(\tau)}^{(\alpha)} \tag{41}$$

$$\breve{W}_{(\tau+1)}^{(\alpha)} = C_{(\tau+1)} e^{-\delta} \sum_{\alpha=1}^{n} \theta^{(\alpha)} W_{(\tau)}^{(\alpha)} \tag{42}$$

where

$$\theta^{(\alpha)} = \frac{1-0.95}{1-n} \tag{43}$$

$$\delta = \frac{1}{2}(Z_{(\tau+1)} - \overline{z}^{(\alpha)})^{T}[F_{z}]^{-1}[Z_{(\tau+1)} - \overline{z}^{(\alpha)}] \tag{44}$$

in which

$$F_{z} = HM_{(\tau+1)} H^{T} + R \tag{45}$$

$$\overline{z}^{(\alpha)} = H\Gamma \widehat{X}_{(\tau)}^{(\alpha)} + H\Gamma q^{(\alpha)} \tag{46}$$

In Eq. (35), $\widehat{X}_{(k+1)}^{(\alpha)}$ is the conditional estimate of $X_{(\tau+1)}$ given that $q_{(\tau+1)} = q^{(\alpha)}$ which is weighted by the probability $W_{(k+1)}^{(\alpha)}$. This probability is obtained from Eqs. (41) and (42), where $\theta^{(\alpha)}$ is the element of Markov transition matrix which according to [23] is taken as (43). $C_{(k+1)}$ is the scale factor, which is determined at each iteration such that $\sum_{\alpha=1}^{n} W_{(\tau+1)}^{(\alpha)} = 1$. To minimize the effect of noise on the weighting terms, a first-order low-pass filter β is embedded and the final weighting equation is obtained as Eq. (41). Initial values for the weighting terms are assumed to be equal i.e. $W_{(0)}^{(\alpha)} = 1$. The covariance matrix Q_b compensates for additional uncertainties such that q may take any value between $q^{(\alpha)}$ and $q^{(\alpha+1)}$. In the above equations, K represents the Kalman gain, P the error covariance of the estimated input vector, Q the process noise covariance, R the measurement noise covariance and Q_b the input covariance.

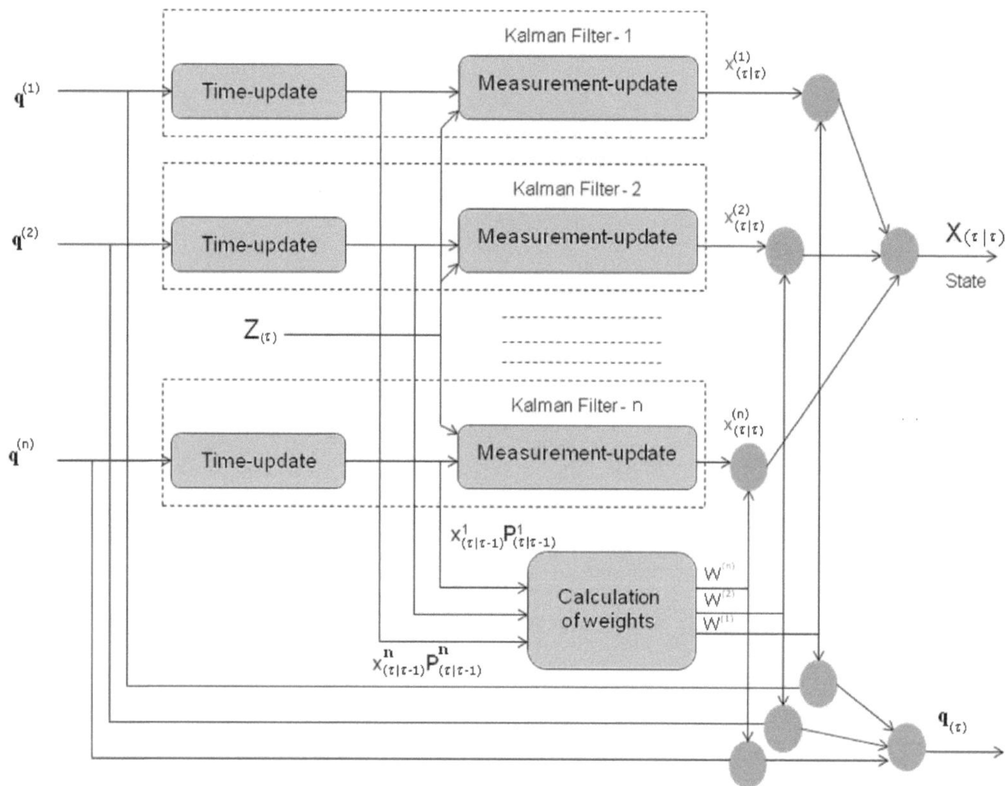

Fig. 7 Block diagram of Adaptive input estimator

Now we can summarize the procedure to estimate the input forces using ASE in the following three steps:

1. Identify a reasonable and justifiable range for the input flux according to the situation.

2. Evaluate the Kalman gain using Eqs. (38) through (40).

3. Get the conditional estimate $\widehat{X}_{\tau+1}^{(\alpha)}$ and hence the estimated state and input flux using Eqs. (35), (36) and (41) through (46).

IV. RESULTS

The cube under consideration is investigated using three types of input fluxes to examine and prove the performance capabilities and deficiencies of the algorithm and its robustness to varying input. The different input fluxes are chosen keeping in mind the practical situations and the insurance of estimator's capability to estimate even the very intricate input fluxes. Hence the three fluxes comprise of sinusoidal input heating the domain, a combination of triangular, rectangular and sinusoidal input heating the system and a step input first increasing and the decreasing step-wise heating the system.

Let us first discuss the estimation of sinusoidal heat flux input. The applied heat flux is given as

$$q(t) = 1 + \sin(\frac{3\pi t}{t_f}),\ 0 \le t \le t_f \tag{47}$$

Δt is taken as 0.01 and total simulation time $t_f = 20$ units. The true flux distribution and the resulting temperature distribution along the x-axis with $(y, z) = (0.5, 0.5)$ and the estimated flux and temperature distribution are presented in Fig. 7. Measurement noise covariance R is taken as 10^{-03} while process noise covariance Q is assumed as 10^{-02} which is a sufficient noise for input estimation problems and not many estimators are able to work in such stringent noise levels [11]. Mesh structure is chosen as $9 \times 9 \times 9$ mesh nodes.

As clear from Fig. 8, the results are quite outstanding especially the state estimation has been excellent. There is a characteristic offset in input flux estimation with respect to both magnitude and time. However such an offset does not undermine the usability and effectiveness of the filter and the results can be regarded as satisfactory and depict the might of ASE in noisy and changing flux regimes. It is of interest to investigate the performance of the estimation filter with different sensor arrangements. As of the above mentioned results, 5 sensors were used as follows:

23

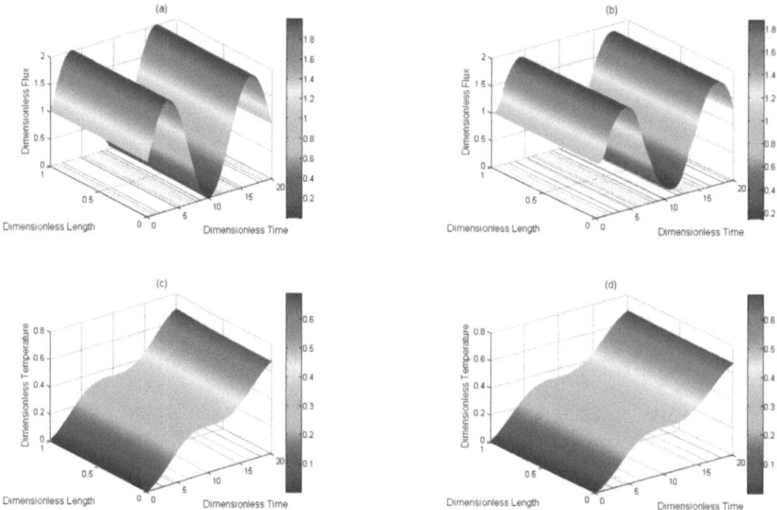

Fig. 8 (a) True flux distribution along the z-axis at $(x, y) = (1, 0.5)$ w.r.t time (b) estimated input flux (c) temperature distribution along x-axis at $(y, z) = (0.5, 0.5)$ w.r.t time (d) estimated temperature distribution, for 5 sensors.

1st sensor: $(x, y, z) = (0, 0.5, 0.5)$

2nd sensor: $(x, y, z) = (0.5, 0.5, 0)$

3rd sensor: $(x, y, z) = (0, 0.5, 1.0)$

4th sensor: $(x, y, z) = (0.5, 0, 0.5)$

5th sensor: $(x, y, z) = (0, 1.0, 0.5)$

The error associated with this sensor arrangement is recorded in Table 2. The error is within the tolerance especially when estimating the temperature distribution. The error is defined as:

$$R_E = \sqrt{\frac{\sum (\mathbb{C}_{true} - \mathbb{C}_{estimated})^2}{\sum (\mathbb{C}_{true})^2}} \tag{48}$$

where $\mathbb{C}$ is the property of interest i.e. flux or temperature. Fig. 9 represent the simulation results with only one sensor installed on the center of the left face i.e. at $(x, y, z) = (0, 0.5, 0.5)$. As expected the results are deteriorated but still presentable and satisfactory. The reason of altered results is that since now the distance between the input and sensor has increased and these are at two extremes in the domain, therefore the time for the information to reach the sensor has increased, that's why the offset between the true input flux and the estimated one has increased and also the magnitude

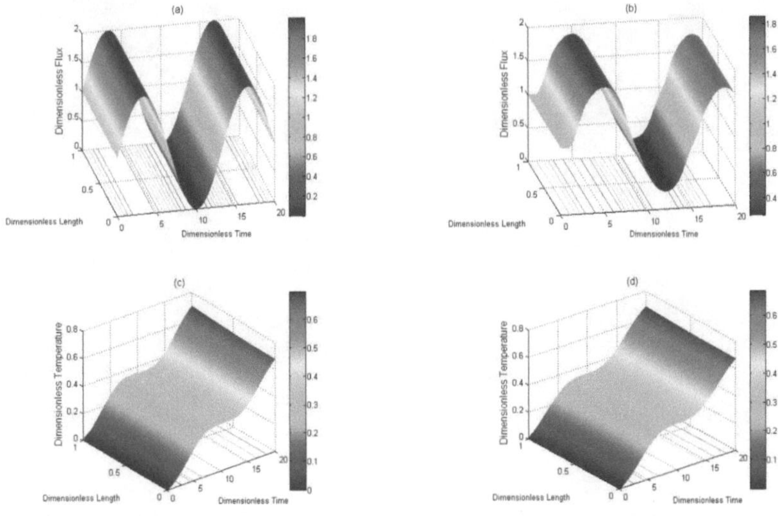

Fig. 9 (a) True flux distribution along the z-axis at $(x,y)=(1,0.5)$ **w.r.t time (b) estimated input flux (c) temperature distribution along x-axis at** $(y,z)=(0.5,0.5)$ **w.r.t time (d) estimated temperature distribution, for 1 sensor.**

estimated is effected. However the state i.e. the temperature distribution is estimated excellently. As given in Table 2, the overall error associated in this scenario has increased especially for input flux.

An optimized performance, with respect to sensor number and error associated, should be expected for 3 sensors located at

$(x,y,z)=(0,0.5,0.5)$
$(x,y,z)=(0.5,0,0.5)$
$(x,y,z)=(0.5,0.5,1.0)$

Fig. 10 represent the estimation results with this sensor arrangement. The results have been considerably improved and as shown in Table 2, the error associated with this sensor arrangement is satisfactory and the estimation performance is established. It is worth mentioning that if somehow it is possible to install the sensors on the surface conducting heat flux in to the domain, the estimation results would be close to perfect.

But since this situation is not practical hence it is ineffectual to conduct analysis with such a sensor arrangement.

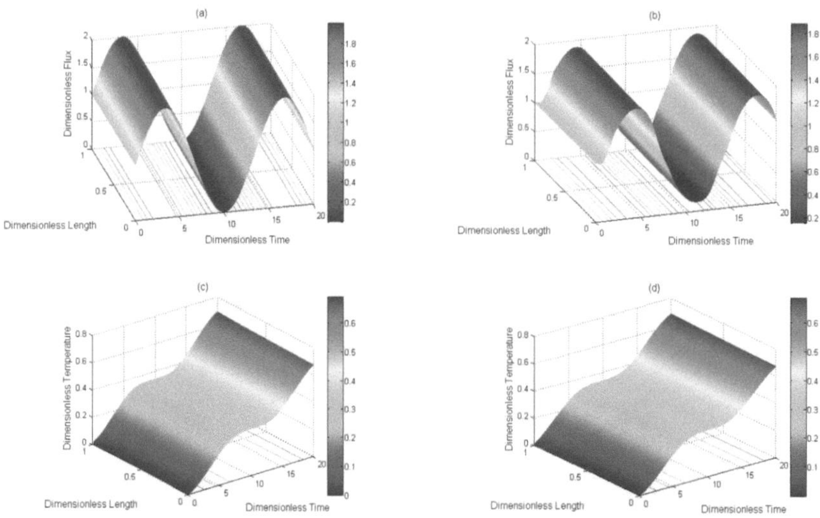

Fig. 10 (a) True flux distribution along the z-axis at $(x, y) = (1, 0.5)$ w.r.t time (b) estimated input flux (c) temperature distribution along x-axis at $(y, z) = (0.5, 0.5)$ w.r.t time (d) estimated temperature distribution, for 3 sensors.

Table 2 Error table for different sensor arrangements for first scenario

	Error in input estimation (%)	Error in state estimation (%)
1 sensor	18.51	1.04
3 sensors	11.37	0.79
5 sensors	9.66	0.33

The inverse estimation filter has proved its worth in the case of a smoothly changing sinusoidal flux. However, sometimes the change of heat flux is not smooth. In this section a scenario has been presented when there is abrupt increase in the input flux, then a linear decrease and then a half sine-wave representation. The scenario poses a tedious job to the ASE algorithm as all the above changes are happening in tandem and the estimator is not given time to stabilize. The applied heat flux can be represented as below:

$$q(t) = \begin{cases} 0 & 0 \le t < 1.0 \\[2ex] 2 & 1.0 \le t < \dfrac{t_f}{2} \\[2ex] 6 - 0.4t & \dfrac{t_f}{2} \le t < \dfrac{t_f}{4} \\[2ex] 1 + 2.25\sin(\dfrac{4\pi t}{t_f}) & \dfrac{t_f}{2} \le t < t_f \end{cases} \tag{49}$$

Again all the three sensor arrangements are analyzed and the associated errors are recorded in Table 3. One thing common in all the three sensor arrangement results is the undefined glitch for the first portion of input i.e. when the input flux is equal to zero. This phenomenon is called the 'transient time' of the filter which actually is the time required for an estimation filter to start estimating the parameter correctly. Same kind of behavior is present in the sinusoidal input case but it is more pronounced in this scenario. Every estimation filter has this deficiency incorporated in it, however the time duration and transient error phenomenon varies algorithm to algorithm. Since this phenomenon is present only for a very small time period, hence it does not preclude the estimator's efficacy. Fig. 11 represent the estimation results for present input scenario.

Notice abrupt change in the flux at time $t = 1.0$ for the estimated values of input flux. This is the point at which some unexpected delay in catching up of stable estimation is prominent for reconstruction results. Since the state and input estimation depend on the single weight matrix, therefore, when there is a sudden and abrupt shift in the values, the estimator needs a little time to converge. Hence, interdependence in estimation of different variables leads to unwanted delay when there are abrupt changes [29]. This lag goes on till the end of estimation and that is why we see the error involved. However there is no such delay in linear and sinusoidal input flux application. The associated error, although higher than that associated with sinusoidal input, is still within the reasonable tolerance level and the estimation performance can be conveniently regarded as satisfactory and excellent state estimation is witnessed in this scenario as well. Fig. 12 and 13 represent the simulation results for 1 sensor and 3 sensors respectively with the identical sensor locations as in the sinusoidal input case for these sensor arrangements. The related error with state and input estimation for these cases is noted in Table 3. As clear from the Figs., a higher error is associated with 1 sensor and the lag is more pronounced than in the case of 5 sensors. However the three sensor arrangement is again the optimized one when the sensor number and error involved is analyzed at the same time.

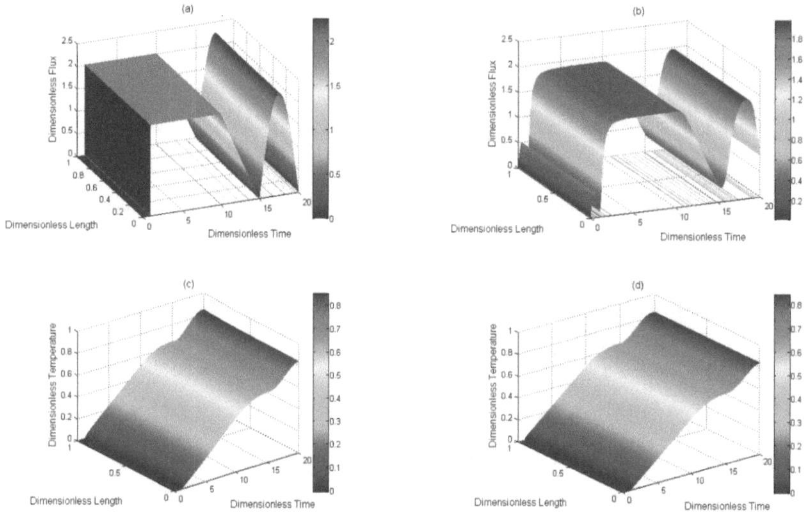

Fig. 11 (a) True flux distribution along the z-axis at $(x,y)=(1,0.5)$ w.r.t time (b) estimated input flux (c) temperature distribution along x-axis at $(y,z)=(0.5,0.5)$ w.r.t time (d) estimated temperature distribution, for 5 sensors.

Table 3 Error table for different sensor arrangements for second scenario

	Error in input estimation (%)	Error in state estimation (%)
1 sensor	18.52	1.0
3 sensors	14.24	0.96
5 sensors	12.45	0.54

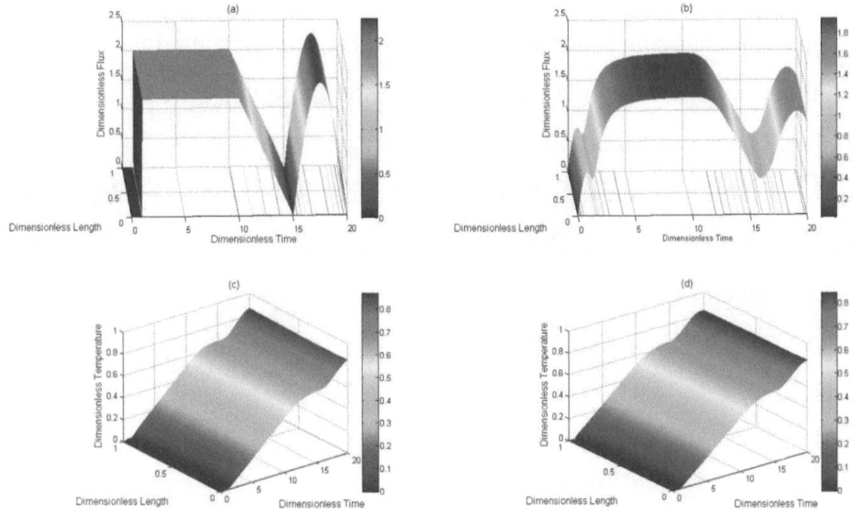

Fig. 12 (a) True flux distribution along the z-axis at $(x,y)=(1,0.5)$ w.r.t time (b) estimated input flux (c) temperature distribution along x-axis at $(y,z)=(0.5,0.5)$ w.r.t time (d) estimated temperature distribution, for 1 sensor.

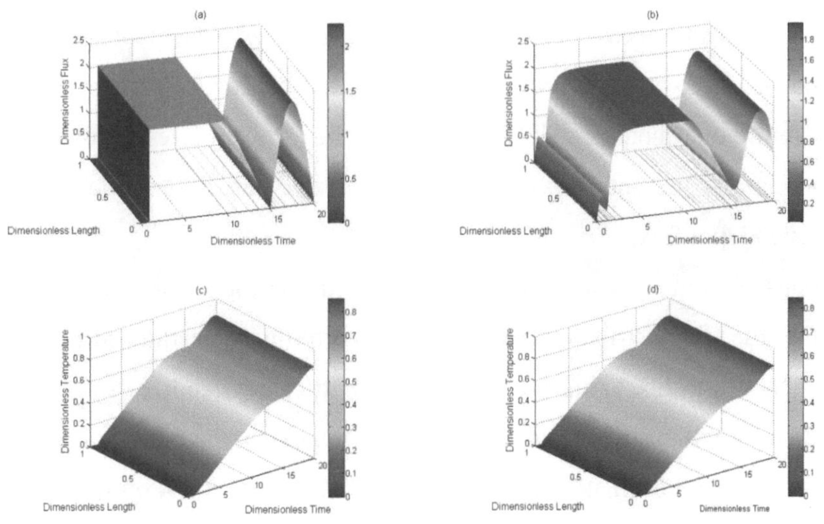

Fig. 13 (a) True flux distribution along the z-axis at $(x,y)=(1,0.5)$ w.r.t time (b) estimated input flux (c) temperature distribution along x-axis at $(y,z)=(0.5,0.5)$ w.r.t time (d) estimated temperature distribution, for 3 sensors.

A more wearying test for the ASE estimation filter can be the one involving continuous step changing input flux. Although this kind of situation is not practically common, but if the estimator is successful in estimating this kind of flux distribution, then the ASE performance will be established and one can easily deduce the wide ranging applicability of the algorithm in theoretical and practical applications as well. This section therefore reports the step-changing input estimation by ASE. The tested input flux is given as:

$$q(t) = \begin{cases} 0.75 & 0.0 \leq t < 3.0, 17.0 \leq t < 20.0 \\ 1.25 & 3.0 \leq t < 6.0, 14.0 \leq t < 17.0 \\ 1.75 & 6.0 \leq t < 9.0, 11.0 \leq t < 14.0 \\ 2.25 & 9.0 \leq t < 11.0 \end{cases} \tag{50}$$

Again all the three sensor arrangements are analyzed. As mentioned earlier, this scenario is a very difficult one from the filtering point of view. The abrupt changes are more pronounced, more frequent and the time between two peaks is very small. As mentioned earlier, the state and input estimation depend on the single weight matrix; therefore, when there is a sudden and abrupt change in the estimation parameter, the estimator needs time to converge and stabilize. Hence, interdependence in estimation of different variables leads to unwanted delay when there are abrupt changes, therefore one should expect more error in the estimation results of this input flux set-up. Fig. 14 presents the estimation results for the 5 sensor arrangement scenario for step input flux.

The transient time phenomenon is there as like before. The sharp edges as present in the true flux distribution are no where visible in the estimated results, the estimator catches the flux value but as soon as it tries to stabilize, since there is another step hence in the struggle of estimating the next step, the estimator presents a wavy look, but over-all estimation performance is satisfactory if not out-standing and the state estimation is excellent as witnessed in other input scenarios. ASE here again has shown its prowess by giving satisfactory and presentable results. The trend is visible and close to the true value and the associated error, as presented in Table 4. is although high but still within the conventional tolerance limits.

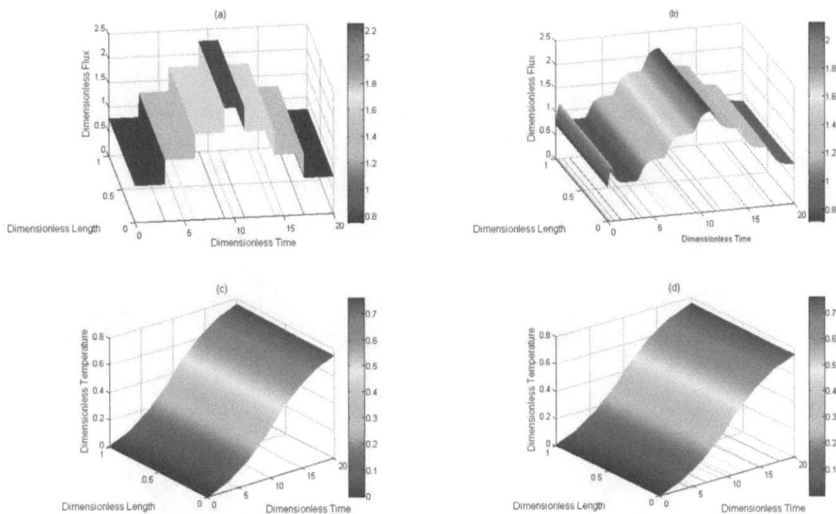

Fig. 14 (a) True flux distribution along the z-axis at $(x,y)=(1,0.5)$ w.r.t time (b) estimated input flux (c) temperature distribution along x-axis at $(y,z)=(0.5,0.5)$ w.r.t time (d) estimated temperature distribution, for 5 sensors.

Fig. 15 and 16 represent the simulation results for 1 sensor and 3 sensors respectively with the identical sensor locations as in the previous cases for these sensor arrangements. The related error with state and input estimation for these cases is noted in Table 4. As clear from the Figs., a very high error is associated with 1 sensor and the estimation performance is imperfect. The step-changing trend is no where visible and a wavy trend is more dominant. The results with three sensors are also not very good but still presentable. The state has been estimated excellently again. Therefore in this case it is recommended to use higher number of sensors that is at least the 5 sensors.

Table 4 Error table for different sensor arrangements for second scenario

	Error in input estimation (%)	Error in state estimation (%)
1 sensor	24.34	0.93
3 sensors	17.57	0.92
5 sensors	12.92	0.55

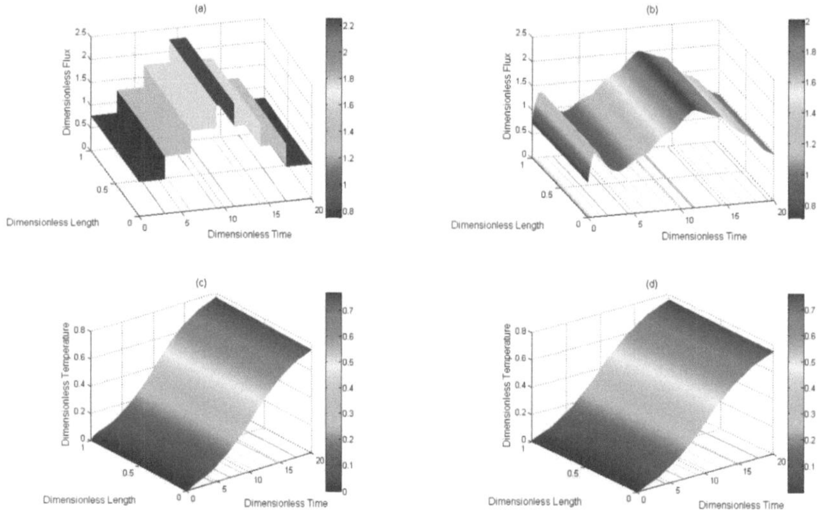

Fig. 15 (a) True flux distribution along the z-axis at $(x, y) = (1, 0.5)$ **w.r.t time (b) estimated input flux (c) temperature distribution along x-axis at** $(y, z) = (0.5, 0.5)$ **w.r.t time (d) estimated temperature distribution, for 1 sensor.**

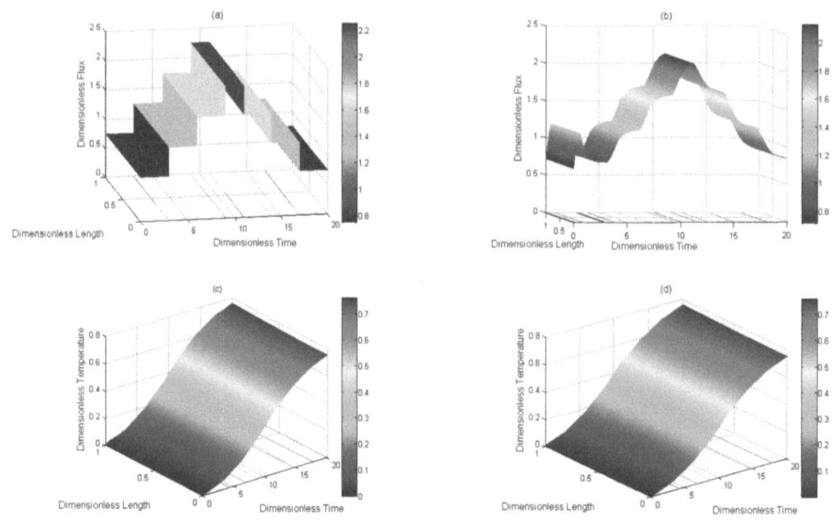

Fig. 16 (a) True flux distribution along the z-axis at $(x, y) = (1, 0.5)$ **w.r.t time (b) estimated input flux (c) temperature distribution along x-axis at** $(y, z) = (0.5, 0.5)$ **w.r.t time (d) estimated temperature distribution, for 3 sensors.**

V. CONCLUSION

An efficient algorithm has been introduced for estimating the unknown boundary input heat flux and the temperature distribution in the inverse heat conduction problem. A direct inverse formulation is constructed using the reverse matrix, which is derived from the governing equations as well as initial and boundary conditions. Three input set-ups have been thoroughly examined to show the robustness of the proposed application of the algorithm. Apart from the first input setup, the other two input scenarios are known to be computationally intensive and intricate as far as the inverse estimation problems are concerned. The inverse estimation algorithm, ASE, has proved its outstanding estimation capabilities in all the scenarios. Different sensor arrangements have been considered and keeping in view the 3-dimensional geometry and process and measurement noise levels, the estimator has proved its prowess and applicability in both theoretical and applications. Although a fairly fine mesh structure was chosen, i.e. *9x9x9* mesh nodes, still the algorithm quite efficiently with respect to simulation time and total analysis time for one scenario was 10 iterations per second on an Intel based dual-core processor computer. However the emerging advances in computing technology and the availability of oct-core processor computers can surely mitigate almost all computational intensities and even very fine mesh structures can be efficiently computed.

REFERENCES

[1] Murio, D. A. and Zheng, H. C. 1995. A Stable Algorithm for 3D-IHCP. *Computers Math. Applic.* 29(5): 97-110.

[2] Zheng, H. and Murio, D. A. 1996. 3D-IHCP on a Finite Cube. *Computers Math. Applic.* 31(1): 1-14.

[3] Scarpa, F. and Milano, G. 1995. Kalman Smoothing Technique Applied to the Inverse Heat Conduction Problem. *Numer. Heat Transfer Part B,* 28: 79-96.

[4] Kaipio, J. and Somersalo, E. 1999. Nonstationary Inverse Problems and State Estimation. *J. Inverse Ill-Posed Problems*, 7: 273-282.

[5] Tuan, P.C. Ji, C.C. Fong, L.W. and Huang, W.T. 1996. An Input Estimation Approach to On-Line Two-Dimensional Inverse Heat Conduction Problems, *Numer. Heat Transfer Part B*, 29: 345–363.

[6] Tuan, P.C. Ji, C.C. Fong, L.W. and Huang, W.T. 1997. Application of Kalman filtering with input estimation technique to on-line cylindrical inverse heat conduction problems. *JSME Int. J. Ser. B*, 40 (1): (1997) 126–133.

[7] Tuan, P.C. Lee, S.C. and Hou, W.T. 1997. An Efficient On-Line Thermal Input Estimation Method Using Kalman Filter and Recursive Least Square Algorithm. *Inverse Problem Eng.*, 5: 309–333.

[8] Tuan, P.C. and Hou, W.T. 1998. The Adaptive Robust Weighting Input Estimation Method for 1-D Inverse Heat Conduction Problem, *Numer. Heat Transfer Part B*, 34: 439–456.

[9] Tuan, P.C. and Ju, M. C. 2001. Adaptive Weighting Input Estimation Algorithm for One Dimensional Cylindrical Heat Conduction Problems, *Proc. Natl. Sci. Counc. ROC(A)*, 25 (3): 163–171.

[10] Tuan, P.C. and Chen, T.C. 2005. Input estimation method including finite element scheme for solving inverse heat conduction problems, Numer. *Heat Transfer Part B*, 47 (3): 277–290.

[11] Ji, C.-C. Tuan, P.C. and Jang, H.-Y. 1997. A Recursive Least-Squares Algorithm for On-Line 1-D Inverse Heat Conduction Estimation, *Int. J. Heat Mass Transfer*, 40 (9): 2081–2096.

[12] Tuan, P.C. Fong, L.W. 1996. An IMM Tracking Algorithm with Input Estimation, *Int. J. Sys. Sci.* 27 (7): 620–639.

[13] Jang, H.-Y. Tuan, P.C. Chen, T. C and Chen, T. S. 2006. Input Estimation Method Combined with the Finite-Element Scheme to Solve IHCP Hollow-

Cylinder Inverse Heat Conduction Problems, *Numer. Heat Transfer Part A*, 50(3): 263–280.

[14] Kalman, R. E. 1960. A New Approach to Linear Filtering and Prediction Problems, *Trans. ASME J. Basic Eng.*, 82(Series D): 35-45.

[15] Park, H.M. and Jung, W.S. 2001. On the solution of Multidimensional Inverse Heat Conduction Problems Using an Efficient Sequential Method, *ASME J. Heat Transfer*, 123: 1021–1029.

[16] Daouas, N. and Rahouani, M.S. 2000. Version Etendue du Filtre de Kalman Discret Appliquee a un Probleme Inverse de Conduction de Chaleur Non-lineaire, *Int. J. Therm. Sci.*, 39: 191–212.

[17] Daouas, N. and Rahouani, M.S. 2004. A New Approach of the Kalman Filter Using Future Temperature Measurements for Nonlinear Inverse Heat Conduction Problems, Numer. *Heat Transfer, Part B* 45: 565–585.

[18] Huang, C.-H. and Tsai, C.-C. 1998. An Inverse Heat Conduction Problem of Estimating Boundary Fluxes in an Irregular Domain with Conjugate Gradient Method, *Heat Mass Transfer*, 45: 47-54.

[19] Ling, X. and Atluri, S. N. 2006. Stability Analysis for Inverse Heat Conduction Problems, *CMES: Computer Modeling in Engineering & Sciences*, 13(3): 219-228.

[20] Hsu, P.-T. 2006. Estimating The Boundary Condition in a 3D Inverse Hyperbolic Heat Conduction Problem, *Appl Math Comput*, 177: 453-464.

[21] Wang, H.M. Chen, T.C. Tuan, P.C. and Den, S.G. 2005. Adaptive-Weighting Input Estimation Approach to Nonlinear Inverse Heat-Conduction Problems, *J. Thermophys. Heat Transfer*, 19: 209–216.

[22] Moose, R. L., Sistanizadeh, M. K., and Skagfjord, G., 1987. Adaptive State Estimation for a System with Unknown Input and Measurement Bias. *IEEE J. Oceanic Eng.*, 12(1): 222-227.

[23] Wang, J. and Zabaras, N. A. 2004. Bayesian Inference Approach to the Stochastic Inverse Heat Conduction Problem, *Int. J. Heat Mass Transfer*, 47: 3927–3941.

[24] Wang, J. and Zabaras, N. A. 2005. Hierarchical Bayesian Models for Inverse Problems in Heat Conduction, *Inverse Probl.*, 21: 183–206.

[25] Wang, J. and Zabaras, N. A. 2005. Using Bayesian Statistics in Estimation of Heat Source in Radiation, *Int. J. Heat Mass Transfer*, 48: 15–29.

[26] Wang, J. and Zabaras, N. A. 2006. A Markov Random Field Model of Contamination Source Identification in Porous Media Flow, *Int. J. Heat Mass Transfer*, 49: 939–950.

[27] Kim, K.Y., Kim, B. S., Kim, H. C., Kim, M. C., Lee, K. J., Chung B. J., and Kim, S., 2003. Inverse Estimation of Time-independent Boundary Heat Flux with an Adaptive Iinput Estimator. *Int. Commun. Heat Mass*, 30(4): 475-484.

[28] Ijaz, U. Z., Khambampati, A. K., Kim, M. C., Kim, S., and Kim, K.Y., 2007. Estimation of Time-Dependent Heat Flux and Measurement Bias in Two Dimensional Inverse Heat Conduction Problems. *Int. J. Heat Mass Tran.*, 50: 4117-4130.

[29] Jazwinski, A. H. 1970. Stochastic Processes and Filtering Theory, Academic Press, New York.

[30] Hornbeck, R. W. 1975. Numerical Methods, Quantum Publishers, New York.

[31] Carslaw, H. S. and Jaeger, J. C. 1959. Conduction of Heat in Solids, Oxford University Press, Amen House, London E.C.4.

[32] Howard, R.A. 1964. System Analysis of Semi-Markov Processes, *IEEE Trans. Mil. Electron*, MIL-8: 114–124.

YOUR KNOWLEDGE HAS VALUE

- We will publish your bachelor's and
  master's thesis, essays and papers

- Your own eBook and book -
  sold worldwide in all relevant shops

- Earn money with each sale

Upload your text at www.GRIN.com
and publish for free